Math
SEEK AND SOLVE

Written and illustrated by Janet Armbrust

Teaching & Learning Company

1204 Buchanan St., P.O. Box 10
Carthage, IL 62321-0010

THIS BOOK BELONGS TO

_____

Cover design by Jenny Morgan

Copyright © 2008, Teaching & Learning Company

ISBN No. 13: 978-1-57310-558-3

Printing No. 987654321

Teaching & Learning Company
1204 Buchanan St., P.O. Box 10
Carthage, IL 62321-0010

The purchase of this book entitles teachers to make copies for use in their individual classrooms only. This book, or any part of it, may not be reproduced in any form for any other purposes without prior written permission from the Teaching & Learning Company. It is strictly prohibited to reproduce any part of this book for an entire school or school district, or for commercial resale. The above permission is exclusive of the cover art, which may not be reproduced.

All rights reserved. Printed in the United States of America.

TABLE OF CONTENTS

Counting	Classroom Counting	4
Directions	Over in the Meadow	6
Sequencing	What Comes Next?	8
Sequencing	Sequencing Safari	10
Measurement	Inching Along	12
Telling Time	Watch the Time!	14
Money	Money for Honey	16
Addition	Alien Addition	18
Addition and Subtraction	Mirror Math	20
Subtraction	Subtracting Snakes	22
Subtraction and Place Value	Seek in the Snow	24
Place Value	Go Fish!	26
Grids	Under the Sea	28
Graphs	Puppies in the Pet Store	30
Multiplication	Seashells by the Seashore	32
Problem Solving Using Multiplication	Colors and Cats	34
Division	Divide Your Wings and Fly	36
Fractions	Funny, Fantastic Fractions	38
Review	The Pond	40
Review	Snowman Fun!	42
	Answer Key	44

Name_____

Classroom Counting

Name_____ COUNTING

Classroom Counting

Let's work on your seeking and counting skills. Can you find and count these items?

1. one elephant

2. two scissors

3. three paintbrushes

4. at least four children

5. at least five cars

6. six snails

7. seven snakes

8. eight hands

9. nine books

10. at least ten crayons

11. at least eleven Ts

12. twelve squares

13. thirteen triangles

14. fourteen diamonds

15. fifteen cherries

16. at least sixteen leaves

17. seventeen flowers

18. eighteen circles

19. nineteen hearts

20. at least twenty stars

TLC10558 Copyright © Teaching & Learning Company, Carthage, IL 62321-0010

Name_____

Over in the Meadow

Name_____

DIRECTIONS

Over in the Meadow

Use your knowledge of positions to help you fill in the blanks.
Above, below, over, under, left and right are words that will help you seek and solve.

1. There are _____ bees <u>above</u> the flowers.

2. There are _____ frogs <u>below</u> the log.

3. There are _____ butterflies <u>over</u> the flowers.

4. There are _____ turtles <u>under</u> the trees.

5. There are _____ ducks on the <u>left</u> side of the meadow.

6. There are _____ deer on the <u>right</u> side of the meadow.

7. How many fish are <u>below</u> the rocks?
   a. 3   b. 4   c. 5

8. How many birds are <u>in</u> and <u>above</u> the trees?
   a. 5   b. 6   c. 7

9. How many grasshoppers are <u>above</u> the log?
   a. 1   b. 3   c. 8

10. How many ladybugs are <u>under</u> the trees?
    a. 4   b. 2   c. 5

―――――――――― Seek and find these items too. ――――――――――

❏ squirrel   ❏ mouse   ❏ numbers 1-9
❏ rabbit   ❏ snake

What Comes Next?

SEQUENCING

What Comes Next?

Fill in the blanks to show what comes next.

1. A B A B ___ ___ ___ ___

2. O X ___ O ___ X O X

3.  ♥ ___ □ ___ △ □ ♥ △

4.  ___ ☆ ◇ ☆ ☆ ___ ☆ ___ ◇ ☆

5. 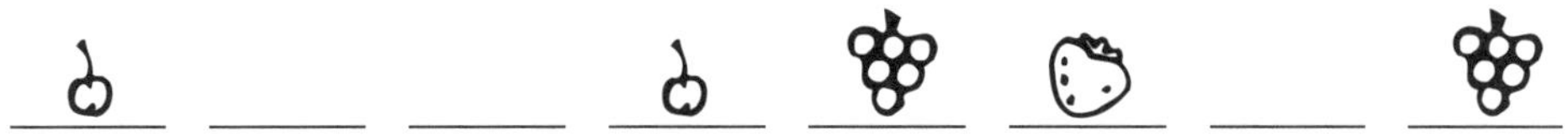 ♪ ___ ___ ♪ 🍇 🍓 ___ 🍇

6.  🍃 ___ ❀ 🍃 ___ ❀ 🍃 🍃 ___

7. 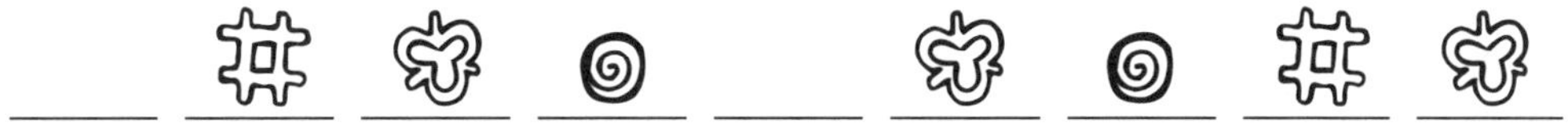 ___ # ❀ ◎ ___ ❀ ◎ # ❀

8. 1 ___ 2 3 1 2 ___ ___ 1

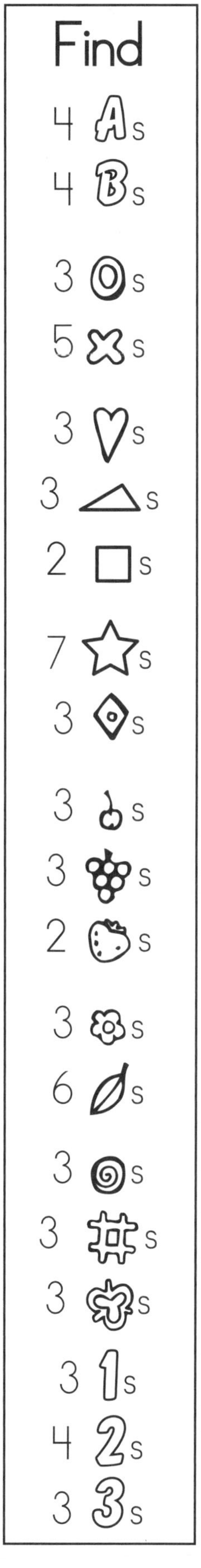

Find

4 A s
4 B s
3 O s
5 X s
3 ♥ s
3 △ s
2 □ s
7 ☆ s
3 ◇ s
3 ♪ s
3 🍇 s
2 🍓 s
3 ❀ s
6 🍃 s
3 ◎ s
3 # s
3 ❀ s
3 1 s
4 2 s
3 3 s

9

Name_____

Sequencing Safari

Name_____

SEQUENCING

Sequencing Safari

Fill in the blanks to show what comes next.

Find
all numbers 1 to 9
all ABCs
12 ☆ s
7 ▫ s
4 ⬡ s
10 ◯ s
5 🐻 s

1. 1   2   __   4   __   6   7   __   __

2. A   B   __   D   E   __   G   H   __

3. 2   ☆   6   8   __   12   __   16   18

4.   __   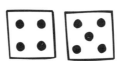   __

5. 3   ⬡   9   ⬡   __   18   21   ⬡   27

6. (spinner: 40, 5, 10, __, 20, 30, __, __)

7. 10   9   8   __   6   __   4   3   __

Name_____

Inching Along

Name_____   **MEASUREMENT**

Inching Along

You will need a ruler for this page.

There are 12 inches in a foot.　　　　　There are 3 feet in 1 yard.
There are 36 inches in 3 feet.　　　　　There are 5,280 feet in 1 mile.

1. Can you find 5 inch worms? How long are they? _____

2. Can you find 14 flowers?
   Measure the tallest one in the pot. How tall is it? _____

3. How wide is the top of the flowerpot? _____

4. How tall is the window? _____

Can you find 5 more objects that are one inch long?

5. _____   8. _____

6. _____   9. _____

7. _____

―――――――――――――――― Seek and find these items too. ――――――――――――――――

❑ one turtle　　　　❑ three lizards　　　　❑ three snails
❑ two birds　　　　❑ three butterflies　　❑ numbers 1-9

Name_____

Watch the Time!

Name_____

Watch the Time!

TELLING TIME

Look at all those clocks and watches.
Some clocks have a face  and hands.

The short hand is the hour hand.

The long hand is the minute hand.

1. Fill in the missing numbers on th clock below.

2. Draw hands on the clock above to show 2:30.

3. Can you find a clock that shows 10:00?
4. Can you find a clock that shows 3:30?
5. Can you find a clock with the minute hand on the 11?
6. Can you find a clock with the hour hand on the 6?
7. There are two clocks that tell 8:00. Can you find them?
8. There are two clocks with no hands. Can you find them?
9. A.M. means hours before noon. Can you find 6 suns?
10. P.M. means hours after noon. Can you find 6 moons? ☾

—————————— Seek and find these items too. ——————————

❏ two watches ❏ 10 hearts ❏ three blind mice
❏ five handprints ✋ ❏ at least 32 stars

Name_____

Money for Honey

Name_____   **MONEY**

Money for Honey

Let's practice your money skills.
Fill in the blanks and then find the list of items to the right.

	Find

1. A penny is worth _____ cents.   10 **1**s

2. A nickel is worth _____ cents.   6 **5**s

3. A dime is worth _____ cents.   10 **¢**s

4. A quarter is worth _____ cents.   4 s

5. A dime equals _____ pennies.   10 s

6. A quarter equals _____ nickels.   5 s

7. It takes 4 quarters to equal a dollar.
   How many quarters are in two dollars? _____   8 s

8. If a jar of honey costs $2.25 and you only have quarters,
   how many quarters would you need? _____   3 🍯s

9. A candy bar costs 80¢. You only have dimes.
   How many dimes do you need? _____   7 🍬s

10. A bag of jelly beans costs $3.38.
    You use 3 dollars, 3 dimes and _____ pennies.   10 🫘s

Name_____

Alien Addition

Alien Addition

Solve the problems. Remember to add the ones column first and then add the tens column.

1.
     16
  + 22

2.   41
  + 53

3. ⭐   75
  + 22

4.   34
  + 14

5.   48
  + 21

6. ⭐   59
  + 30

7.   11
  + 79

8.   35
  + 37

9. ⭐   26
  + 56

10.   225
  + 450

11.   364
  + 524

12. ⭐   885
  + 105

Find

3 s

3 s

3 s

6 ♥ s

6 s

6 ☐ s

numbers 1-9

12 ⭐ s

Name _____

Mirror Math

Mirror Math

ADDITION AND SUBTRACTION

Do you know the sum of two numbers is always the same,
no matter which way the numbers are added?
For example: 4 + 3 = 7 and 3 + 4 = 7
Solve the problems, then find the number of items in the star.

1. 3 + 5 = _____   5 + 3 = _____ − 3 = ☆ butterflies

2. 7 + 8 = _____   8 + 7 = _____ − 8 = ☆ triangles

3. 1 + 9 = _____   9 + 1 = _____ − 9 = ☆ octagon

4. 8 + 12 = _____   12 + 8 = _____ − 10 = ☆ circles

5. 25 + 26 = _____   26 + 25 = _____ − 48 = ☆ number 3s

6. 80 + 20 = _____   20 + 80 = _____ − 99 = ☆ turtle

7. 17 + 9 = _____   9 + 17 = _____ − 17 = ☆ hearts

8. 3 + 15 = _____   15 + 3 = _____ − 13 = ☆ moons

9. 19 + 35 = _____   35 + 19 = _____ − 50 = ☆ birds

10. 60 + 12 = _____   12 + 60 = _____ − 69 = ☆ music notes

Name_____

Subtracting Snakes

22

Name_____

SUBTRACTION

Subtracting Snakes

Seems like subtraction can be fun!
Solve and seek.

1. 28
   -12

2. 54
   -41

3. 37
   -23

4. 79
   -35

5. 92
   -11

6. 86
   -64

7. 45
   -17

8. 63
   -38

9. 72
   -43

10. 734
   -513

11. 898
   -164

12. 630
   -211

Find

1

2 s

3 🧦 s

4 letter S s

5 🌱 s

6 **6** s

7 **7** s

8 ◼ s

9 ❋ s

10 ☆ s

11 🐌 s

12  s

You have already found 6s and 7s.
Now find 1, 2, 3, 4, 5, 8 and 9!

TLC10558 Copyright © Teaching & Learning Company, Carthage, IL 62321-0010

23

Name _____

Seek in the Snow

24

Name_____

SUBTRACTION AND PLACE VALUE

Seek in the Snow

Solve the problems and fill in the blanks.

1. 7
   -4
   ———

Find _____ polar bears.

2. 2
   -1
   ———

Find _____ igloo.

3. 27
   -26
   ———

Find _____ heart.

4. 158
   -154
   ———

Find _____ snowmen.

Enter place values in the snowflakes.

5. 51
   -31
   ———
   20

 tens

Find this many pine trees.

6. 73
   -23
   ———

 tens

Find this many pairs of mittens.

7. 98
   -38
   ———

 tens

Find this many jingle bells.

8. 100
   - 70
   ———

 tens

Find this many hats.

9. 100
   - 0
   ———

 hundreds

Find that many horse and sleigh.

10. 300
    -100
    ———

 hundreds

Find this many ice skaters.

11. 525
    -125
    ———

 hundreds

Find this many scarves.

12. 1081
    - 281
    ———

 hundreds

Find this many reindeer.

13. 1,743
    - 743
    ———

 thousands

Find that many Santa.

14. 9,876
    -2,876
    ———

 thousands

Find this many stars.

15. 4,321
    -2,321
    ———

 thousands

Find this many cups of cocoa.

16. 5,130
    -1,130
    ———

 thousands

Find this many sprigs of holly.

TLC10558 Copyright © Teaching & Learning Company, Carthage, IL 62321-0010

25

Go Fish!

Name_____ PLACE VALUE

Go Fish!

Read the word, write the number, fill in the fish, then find the number of fish described.

1. nine hundred fifty-two = _____

   Find  spotted fish.
   hundreds

   Find  angelfish.
   tens

   Find  star fish.
   ones

2. seven hundred forty-six = _____

   Find  striped fish.
   hundreds

   Find  long fish.
   tens

   Find short fish.
   ones

3. 900 + + 7 = 947
   tens

4. + 70 + 3 = 873
   hundreds

5. + 50 + 2 = 552
   hundreds

TLC10558 Copyright © Teaching & Learning Company, Carthage, IL 62321-0010   27

Under the Sea

Name_____   **GRIDS**

Under the Sea

Use the underwater grid to find sea creatures. Then circle the correct answer.

1. What can you find at A1?
   a. mermaid   b. dolphin   c. sea horses

2. What can you find at B7?
   a. diver   b. submarine   c. sand castle

3. What can you find at H2?
   a. octopus   b. manta ray   c. stripped fish

4. What can you find at C3?
   a. crab   b. jellyfish   c. spotted fish

5. What can you find at D5?
   a. mermaid   b. dolphin   c. sea turtle

6. What can you find at I7?
   a. diver   b. octopus   c. sand castle

7. What can you find at E8?
   a. jellyfish   b. sea horses   c. sea monster

8. What can you find at G6?
   a. shells   b. octopus   c. school of fish

9. What can you find at F4?
   a. submarine   b. sea turtle   c. spotted fish

TLC10558 Copyright © Teaching & Learning Company, Carthage, IL 62321-0010

Name_____

Puppies in the Pet Store

Puppies in the Pet Store

Fill out the charts and graphs as you solve and seek.

Each window has a number. Use the tally chart to help you find the dogs.
Fill in the totals and find them!

	1	2	3	4	Totals
spotted dogs	/				
long-haired dogs		/	//	/	
sleeping dogs	/			/	
tail-wagging dogs	//	//		/	

Can you find these items and fill in the bar graph?

	1	2	3	4	5	6	7	8	9	10
6 bones										
4 dishes										
10 pawprints										
3 tongues										

Study the graph and answer the questions.

- dogs
- kittens
- birds
- rabbits

1. How many dogs can you find? _____
2. How many kittens do you see? _____
3. How many birds can you find? _____
4. How many rabbits do you see? _____

Name_____

Seashells by the Seashore

32   TLC10558 Copyright © Teaching & Learning Company, Carthage, IL 62321-0010

Name_____

MULTIPLICATION

Seashells by the Seashore

Fill in the blanks for the multiplication problems.
Then find the number of items in the balloon.

1. 2 x 5 = ____

2. 4 x 2 = ____

3. 5 x 0 = ____

4. 1 x 1,000 = ____

5. 5 x 6 = ____

6. 7 x 2 = ____

7. 3 x 4 = ____

8. 9 x 1 = ____

9. 8 x 3 = ____

Find ____ balloons.

Find ____ birds.

Find 5 kites.

Find 1 sand castle.

Find 6 children.

Find 7 seashells.

Find ____ hearts. ♡

Find 1 sun.

Find 8 clouds.

TLC10558 Copyright © Teaching & Learning Company, Carthage, IL 62321-0010

33

Colors and Cats

Colors and Cats

PROBLEM SOLVING USING MULTIPLICATION

Solve and seek.

1. Janet has 3 boxes of crayons. Each box has 10 crayons. How many crayons does Janet have? _____ Can you find that many crayons?

2. There are 9 cats. Each cat has 4 legs. How many legs are there in all? _____

3. Janet has 3 cakes and each cake is cut into 8 pieces. How many pieces of cake are there? _____ Janet and her cats ate 7 pieces. How many pieces are left? _____

4. Each cat has 4 toes on each paw. Remember every cat has 4 paws, and there are 2 cats! How many cat toes are there in all? _____

5. That's a lot of toes! Let's count Janet's toes too. She has 5 toes on each foot. She has 2 feet. How many toes are there all together?

    cats' toes _____
    Janet's toes + _____
    total toes = _____

Find
3 _s
9 _s
7 _s
10 _s
2 _s

─── Seek and find these items too. ───

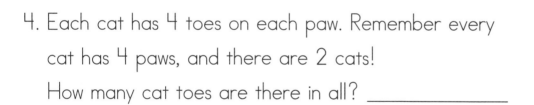

❑ sun   ❑ butterfly   ❑ numbers 1-9
❑ tulip   ❑ turtle

Name_____

Divide Your Wings and Fly

36

Name_____   **DIVISION**

Divide Your Wings and Fly

Both problems in each row have the same answer.
Circle the correct answer, then use the key to see
how many items you need to find.

Key
a. = 1   b. = 2
c. = 3   d. = 4

1. 3)12   4)16   There are _____ s to find.

    a. 6   b. 8   c. 4   d. 2

2. 2)10   3)15   There is _____ , and a to find.

    a. 5   b. 4   c. 2   d. 6

3. 2)14   3)21   There are _____ ☺s to find.

    a. 6   b. 7   c. 5   d. 8

4. 9)81   7)63   There are _____ s to find.

    a. 8   b. 6   c. 12   d. 9

5. 7)42   11)66   There is _____ UR SMART to find.

    a. 6   b. 8   c. 12   d. 9

———————————— Seek and find these items too. ————————————

❏ two butterflies that are the same   ❏ two of each number 1-9

TLC10558 Copyright © Teaching & Learning Company, Carthage, IL 62321-0010   37

Funny Fantastic Fractions

Name_____ **FRACTIONS**

Funny, Fantastic Fractions

It's all clowning around on this page! Write the fractions and seek the items.

☆ Fractions can be used to describe parts of a group.
For example, there are 6 clowns. Three are tall.
The fraction of clowns that are tall is $\frac{3}{6}$ $\frac{\text{tall clowns}}{\text{clowns in the group}}$

☆ The top number is called the numerator.

☆ The bottom number is called the denominator.

1. There are six clowns. Five of them have hats.
   Write the fraction for this. _____

2. Two of the ten hats have flowers on them.
   Write the fraction for this. _____

☆ Fractions can also be used to describe parts of a whole.
This one circle is in two parts. $\frac{1}{2}$ $\frac{\text{part shaded}}{\text{parts of the circle}}$

Look at each picture and write a fraction for the shaded part in each.

Find

3 tall clowns

10 s

10 🌸s

9 triangles

7 squares

4 peace signs

3.  ____   4. 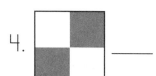 ____   5. 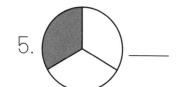 ____

———————— Seek and find these items too. ————————

❑ 10 stars   ❑ numbers 1-9

39

Name _____

The Pond

40

Name_____

REVIEW

The Pond

Solve and seek.

1. Nine ducks are on the pond. Two ducks fly away.
   How many are left? _____ Find that many ducks.

2. A raccoon finds seven snails. He eats five of them.
   How many are left? _____
   Now add six to your answer and find that many snails.

3. 51
   + 12

4. 15
   − 10

5. 96
   + 8

   −100

   Find that number.

   Find that many dragonflies.

   Find that many frogs.

6. In the number 475, which digit has the <u>least</u> value? _____
   Explain:_____
   Find that many fish.

7. 99 ÷ 3 ÷ 11 = ____   48 ÷ 6 ÷ 2 = ____   36 ÷ 9 ÷ 2 = ____

   Find that many flowers.

   Find that many turtles.

   Find that many alligators.

Name _____

Snowman Fun!

Name_____   REVIEW

Snowman Fun!

Use the bar graph to answer each question.

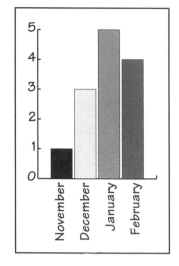

1. How much total snow fell during December and January? _____
   Find that many snowflakes.

2. How much snow fell in all four months combined?
   _____ Find that many mittens.

3. Which month had three times as much snow as November?
   _____ Find three snowmen plus two.

4. There are 8 hats and 7 scarves. How many hats and scarves are there?
   _____ Find them all.

5. There are 8 <u>pairs</u> of mittens in the picture. That makes _____ mittens.
   You have already found _____.
   How many more do you need to find? _____ Can you find them?

6. $2\overline{)16}$ ← Find this many boots.

7. $3\overline{)21}$ ← Find this many birds.

8. There are six buttons in the picture. One of them is on a hat.
   Write the fraction for this. _____ Find the six buttons.

9. Write the fraction for this picture. 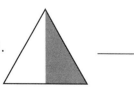 _____
   Can you find 10 triangles?

Answer Key

Classroom Counting (page 4)

Over in the Meadow (page 6)

1. 5
2. 4
3. 5
4. 2
5. 4
6. 3
7. a
8. a
9. b
10. c

What Comes Next? (page 8)

1. A, B, A, B
2. X, X
3. △ ♡
4. ☆ ◇ ☆
5. 🍇 🍪 🍒
6. 🍃 🍃 ❀
7. 🌀 ✳
8. 2, 2, 3

Sequencing Safari (page 10)

1. 3, 5, 8, 9
2. C, F, I
3. 4, 10, 14
4. ⚃ ⚅
5. 12, 15, 24
6. 15, 25, 35
7. 7, 5, 2, 1

Answer Key

...ching Along (page 12)

1. 1 in.
2. 6 in.
3. 6 in.
4. 7 in.
5. bird
6. snail
7. turtle
8. lizard
9. bee

Watch the Time! (page 14)

1. 1, 3, 5, 8, 10

...oney for Honey (page 16)

. 1
. 5
. 10
. 25
. 10
. 5
. 8
. 9
. 8
. 8

Alien Addition (page 18)

1. 38
2. 94
3. 97
4. 48
5. 69
6. 89
7. 90
8. 72
9. 82
10. 675
11. 888
12. 990

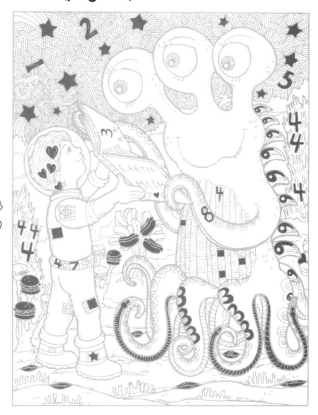

Answer Key

Mirror Math (page 20)
1. 8, 8, 5
2. 15, 15, 7
3. 10, 10, 1
4. 20, 20, 10
5. 51, 51, 3
6. 100, 100, 1
7. 26, 26, 9
8. 18, 18, 5
9. 54, 54, 4
10. 72, 72, 3

Subtracting Snakes (page 22)
1. 16
2. 13
3. 14
4. 44
5. 81
6. 22
7. 28
8. 25
9. 29
10. 221
11. 734
12. 419

Seek in the Snow (page 24)
1. 3
2. 1
3. 1
4. 4
5. 20, 2
6. 50, 5
7. 60, 6
8. 30, 3
9. 100, 1
10. 200, 2
11. 400, 4
12. 800, 8
13. 1000, 1
14. 7000, 7
15. 2000, 2
16. 4000, 4

Go Fish! (page 26)
1. 952, 9, 5, 2
2. 746, 7, 4, 6
3. 4
4. 8
5. 5

Answer Key

Under the Sea (page 28)

1. b
2. a
3. b
4. b
5. a
6. c
7. c
8. b
9. a

Puppies in the Pet Store (page 30)

1. 20
2. 10
3. 5
4. 5

Seashells by the Seashore (page 32)

1. 10
2. 8
3. 0
4. 1000
5. 30
6. 14
7. 12
8. 9
9. 24

Colors and Cats (page 34)

1. 30
2. 36
3. 24, 17
4. 32
5. 32 + 10 = 42

Answer Key

Divide Your Wings and Fly (page 36)
1. c., 3
2. a., 1
3. b., 2
4. d., 4
5. a., 1

Funny, Fantastic Fractions (page 38)
1. 5/6
2. 2/10
3. 1/2
4. 2/4
5. 1/3

The Pond (page 40)
1. 7
2. 2 + 6 = 8
3. 63
4. 5
5. 104, 4
6. 5
7. 3, 4, 2

Snowman Fun! (page 42)
1. 8
2. 13
3. December
4. 15
5. 16, 13, 3
6. 8
7. 7
8. 1/6
9. 1/2